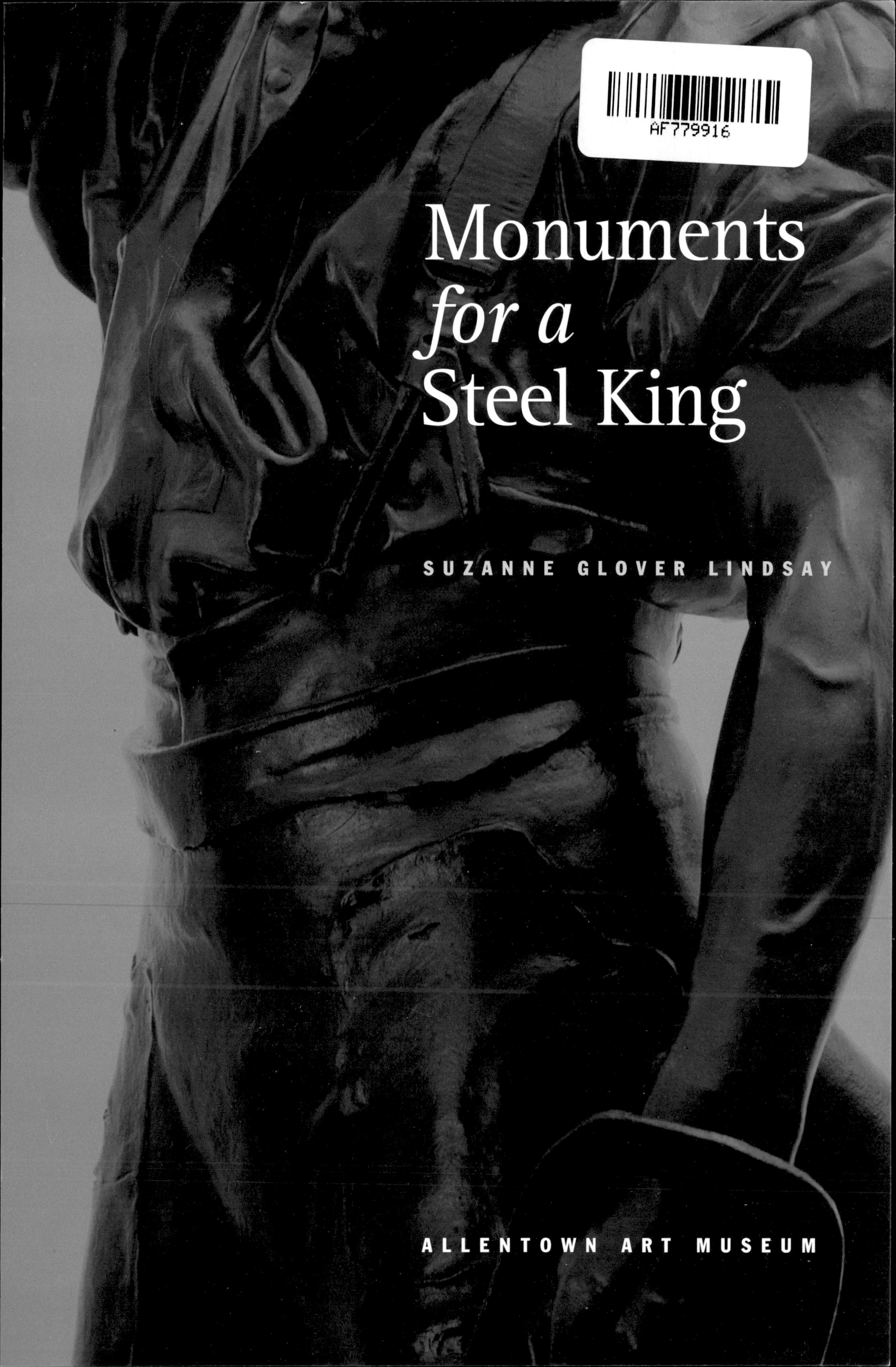

Monuments *for a* Steel King

SUZANNE GLOVER LINDSAY

ALLENTOWN ART MUSEUM

FOREWORD

The Lehigh Valley yielded a number of astonishing surprises during my tenure at the Allentown Art Museum. None had the force of personality of two monumental figures that the great French painter and sculptor Jean-Léon Gérôme made for Charles M. Schwab, the founder and chairman of Bethlehem Steel. In 1981 officials at Bethlehem Steel invited me to inspect a collection of archival material held in a warehouse then in the process of being vacated. On that memorable day I inspected scores of portraits of former Steel executives. I also had the uncanny good fortune to find the palm fronds that had once been an integral part of a reduction of Augustus Saint-Gaudens's *Victory* figure for the *Sherman Monument.* I had seen the sculpture, *sans* palm fronds, in a gardener's shed on the Merle-Smith campus of Moravian Academy some time before. The palm fronds and *Victory* have since, happily, been reunited.

I discovered the Gérôme figures outdoors, forlorn and forgotten. The *Puddler* stood erect on his marble block. *Scientific Metallurgy* was cruelly lashed with rusting iron wires in a horizontal position to a wooden pallet, apparently since 1947. The signatures and foundry marks from the Siot foundry in Paris were clearly legible. Many important sculptural monuments were similarly neglected in the mid-twentieth century; the fashion for monumental figurative sculpture was then perhaps at its low point. Nevertheless, the superior quality of Gérôme's casts was unmistakable.

Clearly, Bethlehem Steel had acquired the figures with the intention of a memorial to its chairman, the late Charles Schwab. My suggestion that they be erected in the city of Bethlehem did not find a sympathetic ear. My second suggestion, that the sculptures be restored to public view at the Allentown Art Museum, elicited a favorable reaction from Eugene Kline, then vice-president for public affairs at Bethlehem Steel, and Howard Feldman, then employed by Steel and a museum trustee. I am grateful to them both for facilitating this magnificent gift. Bethlehem Steel has been an important community partner in many civic and cultural endeavors in the Lehigh Valley. Through its generosity the Gérômes are shared once again with a larger public, as they had been in their original location on the terrace of Schwab's mansion on New York's Riverside Drive.

Bethlehem Steel has long been a dynamic force in the greater Lehigh Valley. Many Valley institutions owe an enormous debt of gratitude to direct corporate or indirect private patronage from Steel, its executives and workforce. To name only a few examples, there is the beautiful collection of paintings that is the core of Lehigh University's art collection, received as a bequest from Steel chairman Eugene Grace. The nucleus of buildings now known as Historic Bethlehem, Inc. was rescued from oblivion by an influential group of Steel executives. And it is well known that Charles Schwab was president and champion of the world-famous Bach Choir of Bethlehem. Schwab as a patron of American and French sculptors has not been published heretofore. Part of that story is the purpose of Suzanne Lindsay's essay.

I first came to know Dr. Lindsay's work some years ago through a memorable exhibition on painter Mary Cassatt she organized for the Philadelphia Museum of Art. More recently she authored the catalogue of nineteenth-century French sculpture in the collection of the National Gallery of Art. Her current project is a comprehensive study of the Degas waxes at the National Gallery.

Lindsay speaks here not only of the formal issues of nineteenth-century sculpture, but of the era in which it was made. She has given us a searching essay on Gérôme's last monumental work, to my knowledge his only large-scale work in the United States. I am grateful for her embrace of the project; she writes with the vigor and insight of an art historian who still has the strong hands of a sculptor.

Peter F. Blume

ACKNOWLEDGMENTS

It is a pleasure for me, as author, to recognize the many individuals who made this project possible from beginning to end. First and foremost, I would like to thank Peter F. Blume for inviting me, while director of the Allentown Art Museum, to examine and write about these extraordinary bronzes. The resulting essay owes much to his research on the pair, while he was curator, at the time of their acquisition. I am especially grateful to Dolores O'Mara, Sofia Bakis, and Carl Schafer who were wonderful in answering questions, making material in their files available, and getting the bronzes photographed, especially during a final round on an uncomfortably hot summer day. Many scholars and professionals contributed to the research. I thank Lance E. Metz, historian at the National Canal Museum, Easton, Pennsylvania, for wide-ranging discussions and Thomas A. Smith, the Canal Museum's Collections Manager/Archivist, for generous help. For information, photographs, and sheer enthusiasm, I owe much to colleagues Michele H. Bogart, Melissa Dabakis, Sarah Bradford Landau, Anne Pingeot, and Edye Weissler. I am also grateful to the many librarians, archivists, and curators who generously provided information by e-mail. Warm thanks to Gary Millenbruch and Bette Kovach of Bethlehem Steel Corporation for their comments on the manuscript. I applaud editor Elaine Koss and designers Natalie Salvucci and Phil Unetic for their professionalism and imagination; the process was as pleasurable as the result. Finally, I am indebted to Christine I. Oaklander for her support and energy in bringing this project to fruition. This handsome and authoritative catalogue does great honor to Schwab and Gérôme, and I am delighted to be part of such a tribute.

Suzanne Glover Lindsay

Jean-Léon Gérôme's *Metallurgical Science* and *Metallurgical Worker* have been on proud display at the Allentown Art Museum since their 1982 donation by Bethlehem Steel. These majestic bronzes, unique in the artist's oeuvre and the last major works he completed, have never before been the subject of a scholarly publication. The Museum is proud to unveil this permanent record of two stellar works from its world-class collection.

Successful completion of this important project would have been impossible without the assistance of Museum colleagues Carl Schafer, Sofia Bakis, and Shana Leigh Herb; photographer Angelo Caggiano; editor Elaine Koss; designers Natalie Salvucci and Phil Unetic; and the array of scholars, curators, archivists, and librarians listed above by Suzanne Lindsay. Tremendous credit goes to former Museum director Peter F. Blume, who was solely responsible for coaxing Bethlehem Steel to part with the Gérôme sculptures, conducted preliminary research on them, and persuaded Dr. Lindsay to craft her fascinating essay. This publication is a fitting climax to Blume's years of hard work and his many accomplishments as curator and director of the Allentown Art Museum.

Christine I. Oaklander, Ph.D.
Curator of Collections and Exhibitions

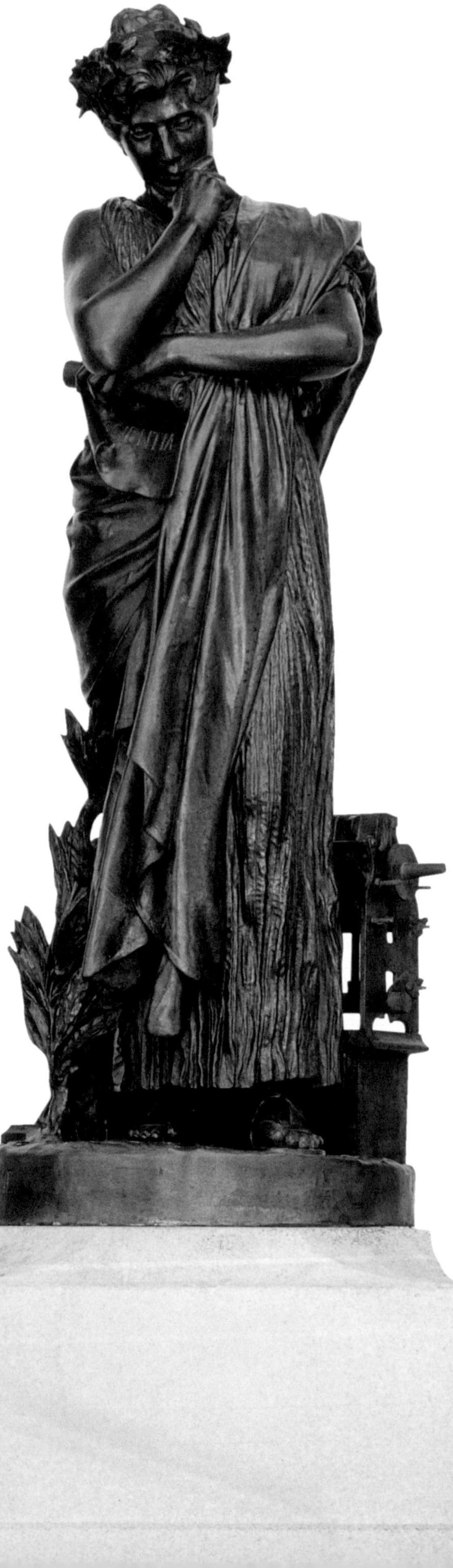

FIG. 1.

Jean-Léon Gérôme, *Metallurgical Science*, 1902-3, bronze, 87 x 35 x 32 in. (without base), Allentown Art Museum, gift of Bethlehem Steel Corporation, 1982.8.1

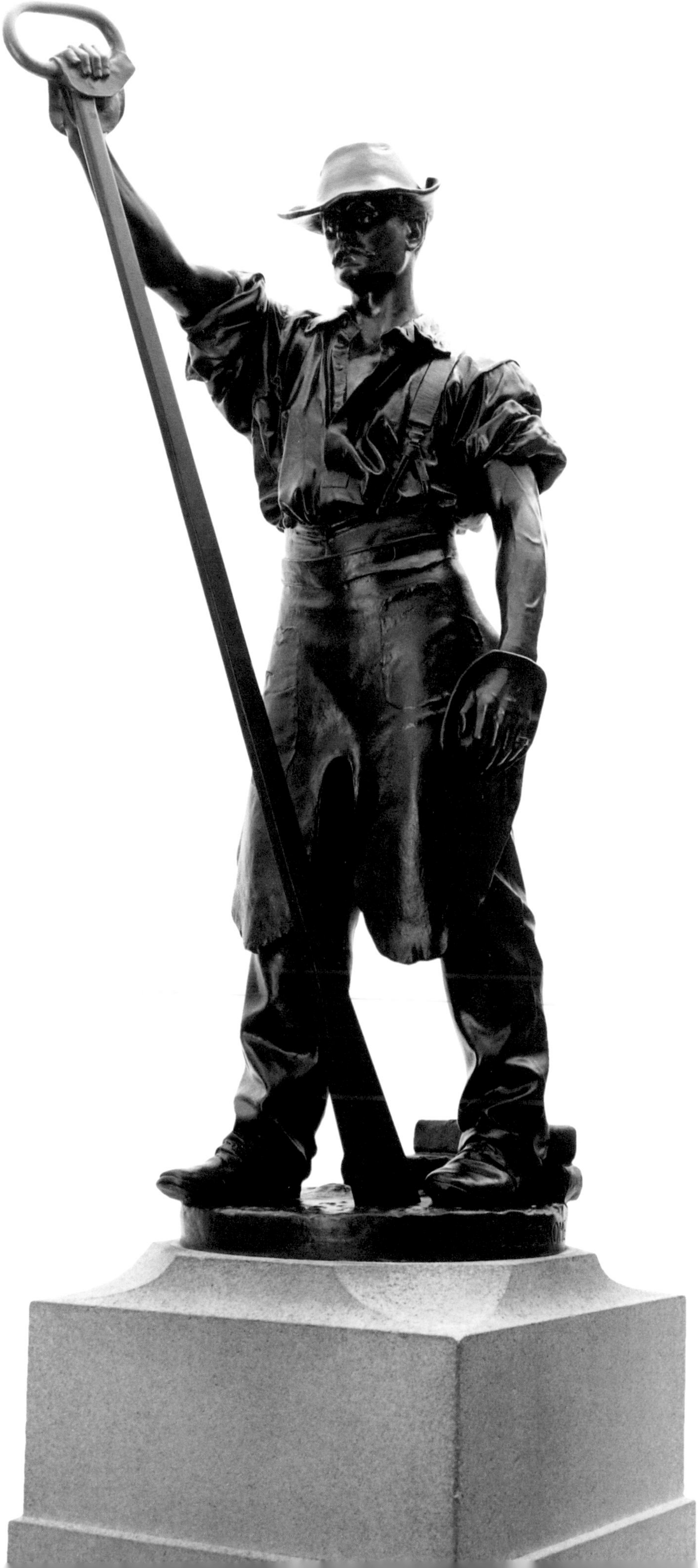

FIG. 2.

Jean-Léon Gérôme, *Metallurgical Worker ("The Puddler")*, 1902-3, bronze, 95 x 31½ x 36½ in. (without base), Allentown Art Museum, gift of Bethlehem Steel Corporation, 1982.8.2

Just outside the Allentown Art Museum stand two imposing bronze statues. To the front, on the Fifth Street porch, the effigy of a meditative woman in classical dress, *Metallurgical Science,* ponders those who pass by (Fig. 1). To the rear, gazing into the distance from his pedestal on the Butz Terrace, is the jaunty figure of a modern American artisan entitled *Metallurgical Worker ("The Puddler")* (Fig. 2). Though they seem very different, they were originally conceived and shown as a pair to convey a multilayered statement about American steelmaking as it entered its heyday.

The two bronzes are important artistically. Completed in 1903, they were the swan song of one of France's most prominent realist artists of the late nineteenth century, Jean-Léon Gérôme (1824–1904). They are unique works: No other casts or versions and only one surviving preliminary sketch are currently known. The statues are also compelling as monumental sculpture, with an arresting physical impact that goes beyond their sheer scale (about seven and eight feet, respectively, without their thirty-one-inch bases) and a lifelike presence dramatized by bold contours and internal rhythms. They are, moreover, excellent casts. They register Gérôme's fluid modeling in subtly gritty bronze that is exceptionally free of flaws and sensitively tooled by the craftsmen at Gérôme's preferred foundry, Siot Fondeur in Paris.[1] Such technical refinement suggests they were made for scrutiny at close range as well as from afar.

These special figures are nonetheless not widely known today, in part because they belonged to private collectors. They left Paris for the New York mansion of their patron, Charles M. Schwab, one of the most colorful and influential steel magnates of the early twentieth century. When the house was torn down in 1947, the bronzes moved to Bethlehem, Pennsylvania, purchased by Bethlehem Steel Corporation, whose preeminence by the 1940s owed much to Schwab, its founder, first president, and chairman of the board. Steel's gift of the two statues to the nearby Allentown Art Museum in 1982 made these unique works of art available to the public as never before.[2] Using Schwab's own library, which Steel divided and gave to various archives around the same time, we can now offer a preliminary account of this unusual project, one that interweaves the art worlds of Paris and New York with the international industrial world dominated increasingly by the United States.

CHARLES SCHWAB: A VERY HUMAN AMERICAN HERO

Charles M. Schwab (1862-1939, Fig. 3) counts among the corporate greats of his day who embodied Horatio Alger's American ideal.[3] He was an individual of modest birth who prospered—by his own account—through sheer ambition, hard work, luck, and the help of some powerful people. A tireless promoter of that path to success, Schwab actively shaped it by highlighting and even altering certain aspects of his life in the countless interviews, profiles, and even "how-to" brochures published about him at the peak of his career.[4] The myth about this steel man eventually blurred and transformed the reality.

He was born Charles Michael Schwab, youngest son of second-generation immigrants who settled in a remote area of central Pennsylvania to be with their own kind, Catholic Germans living in or near the town of Loretto.[5] His father, often portrayed as a textile mill hand whose enterprises failed where the son's succeeded, actually ran a small woolen mill started by his own father and then branched out into other enterprises, including a livery stable and mail coach.[6] The family lived simply but apparently was not poor, as legend would have it.[7] Supported by his family's insistence on education, Schwab pursued some high school courses at nearby St. Francis College—and then went to work. He began as a clerk in a grocery store in nearby Braddock, where he displayed mathematical skills that caught the eye of the superintendent of the nearby Edgar Thomson Steel Works, William R. (Capt. Bill)

FIG. 3.

Photograph of Schwab (bottom center), President Woodrow Wilson, and champion riveters at Hog's Island, PA, Aug. 1918, Bethlehem Steel Collection, National Canal Museum, Easton, PA.

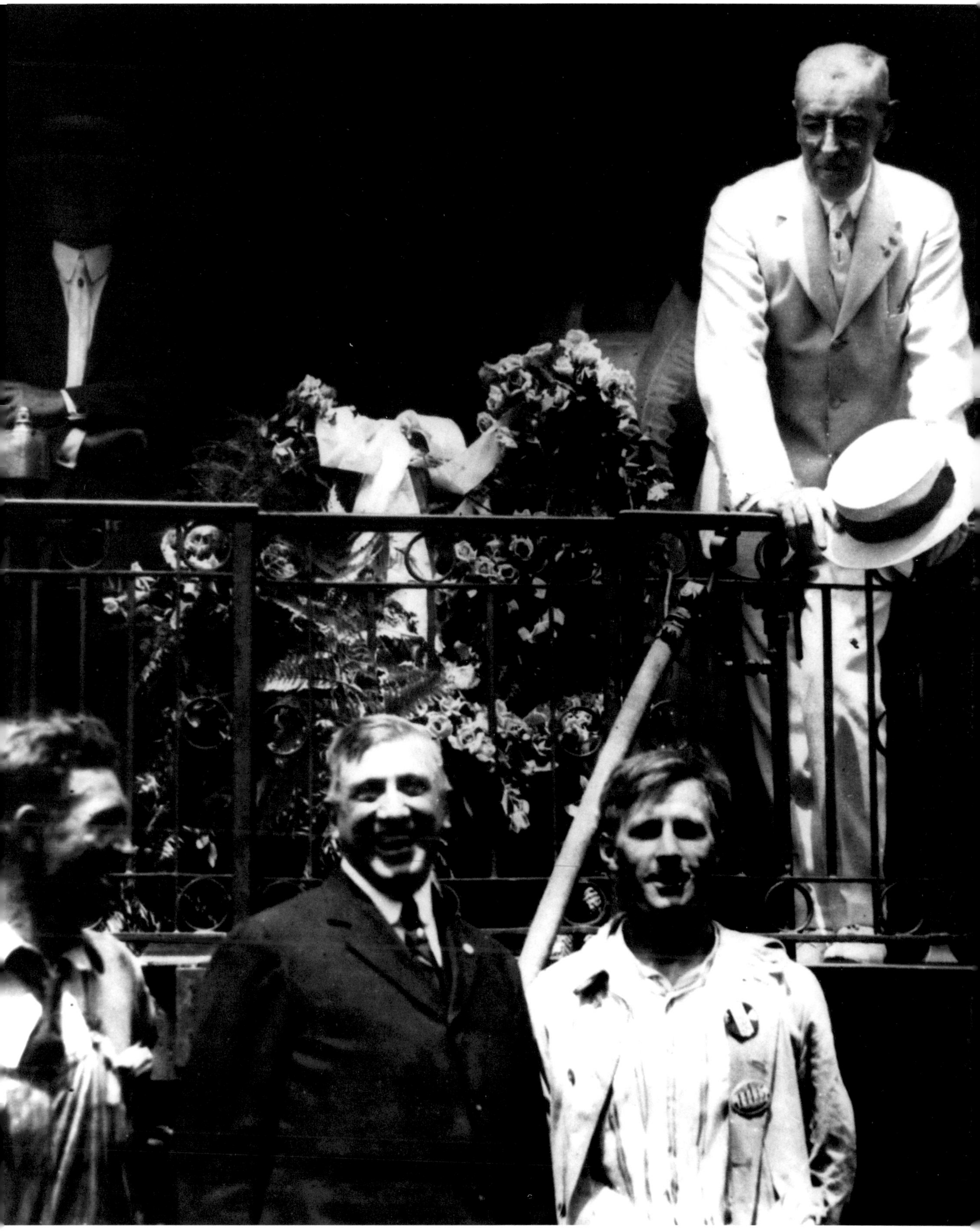

Jones, who promptly hired him. Placed in the engineering corps, Schwab was briefly a surveyor's rodman before landing a draftsman's job. He rose through the ranks and garnered some of the department's most challenging assignments by setting himself apart from his peers through working long hours, often overtime and without pay. The most important individual to notice him was the power behind the company, another Horatio Alger figure, Andrew Carnegie, who soon promoted Schwab to general superintendent of the Homestead plant outside Pittsburgh. He was so successful that Carnegie made him general superintendent at Thomson when Jones was killed on the job in 1889. That appointment put a twenty-seven-year-old in charge of America's largest steelworks after only ten years in the business.[8]

The combination of Carnegie's favor (the two were like father and son) and the opportunities created by the enormous growth of the steel industry in the 1880s and 1890s speedily carried the young man to extraordinary professional heights. The chronology of that rise alone is impressive. By 1897 Carnegie had made Schwab president of Carnegie Steel at thirty-five years of age. Four years later Schwab became president of a new holding company, the United States Steel Company, whose creation he spearheaded through a steel trust that merged Carnegie Steel, on Carnegie's retirement, with nine smaller companies.[9] Headquartered in the corporate capital of the United States, New York City, this novel entity marked the steel industry as the new backbone of American industry and itself, the so-called "Billion Dollar" corporation, as the largest and most expensive in the country. Only thirty-nine at the time, Schwab became one of the most powerful international figures in the private sector. The American Iron and Steel Association (AISA) journal wrote with breathtaking hyperbole on the subject: "Mr. Schwab occupies today the most prominent and the most responsible position in connection with the world's industrial development that has ever been held by any man."[10]

In 1903, however, Schwab retired as president of U.S. Steel with a controlling interest in the small but profitable Bethlehem Steel Company in South Bethlehem, Pennsylvania. This became the primary property of a holding company established the following year in New Jersey, the Bethlehem Steel Corporation, with Schwab as a director and its chairman and president. He immediately whittled, expanded, and diversified operations, and Steel grew large and powerful. In 1918, he went into semiretirement, staying on as chairman of the board, taking charge of the United States Shipping Board Emergency Fleet Corporation and dramatically boosting military shipbuilding in World War I, helping to make the United States a dominant international military force. In 1927 he became president of the American Iron and Steel Association, serving as spokesman for the industry to the government and the world in his final years as a "retired steel manufacturer"–though still the titular chairman of the company that he had founded, "my own child, the soul of my life."[11] His death at seventy-seven in September 1939 made the front page of major American newspapers. He received the ritual honors of a truly eminent citizen, requiem high mass at the largest and most prestigious Catholic church in the country, New York's St. Patrick's Cathedral.[12]

Luck and timing alone did not create Schwab, however. As he often declared, much of his professional success–and his impact as a modern American legend–came from his personal qualities.[13] He was imaginative and intellectually curious, read voraciously, and studied a wide array of topics outside his profession merely for the sake of knowledge. He accomplished much, he said, by being a "dreamer."[14] His improvements in technical and management areas indeed helped his rise to the top. However, his most famous strength was his manner with people. Observers praised him for "executive and managerial greatness, if not genius," thanks to an effective blend of attributes: "Always optimistic, a plunger and a patriot, a hale fellow well met, two-fisted and hard-headed, yet sentimental and friendly, he knew how to master men and use machinery."[15] He was at ease with all social levels in the industry:

"In his bluff, ever youthful way, he regarded them all as 'the boys.' That was true of the executives and of the tens of thousands of sweating, grimy laborers who dealt with fire and flame, ore and molted metals, grails and girders and armor plates–the sinews of industrial America."[16] He influenced through charisma rather than compelling argument. Schwab was handsome and big and exuded vitality; he was blessed with a commanding presence and a positive energy that won over even hostile audiences. His celebrated speaking style perfectly demonstrated those qualities. He persuaded through sheer magnetism, riveting gestures, and a hypnotic honeyed voice, though his actual words were mostly platitudes when they made any sense at all, which was rare: He was often teased for thoroughly scrambled sentences. Schwab relied so heavily on charm that he made its cultivation his seventh "Success Commandment" for others.[17]

Schwab was famous for personal values that success did not spoil. He repeatedly declared that he preferred the simple life even within his vast mansions and remained deeply attached to his roots, his friends, and especially his family. Though immensely wealthy, he alleged he cared little about money for himself, and instead spent extravagantly on his business, home region, intimates, and even strangers in distress.[18] He restored churches, built a convent for his sister, a Discalced Carmelite, a house for his mother, a school for his wife's birthplace (Weatherly, Pennsylvania), and churches and industrial schools at Braddock and Homestead. He regularly returned home in two ways: symbolically, through his private railroad car named "Loretto," and through the elaborate estate that he built near the town itself, "Immergrün" (German for evergreen). He married a "girl from back home," Emma Eurania (Rana) Dinkey. Childless, the couple was portrayed as unusually close and mutual admirers.[19] She called him "Solomon" and he claimed to rely on her judgment in all things. Both were involved in designing and furnishing their homes.[20] Despite her disabling rheumatism and oft-mentioned "queenly" build, she regularly joined Schwab at public functions and was frequently interviewed, especially during the war.[21] According to the obituaries, her death in January 1939 profoundly upset him; his own demise, after years of failing health, followed only months later.

These qualities in an American tycoon reportedly startled the corporate world when Schwab entered the limelight with U.S. Steel. His most familiar champion, Eugene Grace, reported that he "burst upon the New York scene like a comet. He was a figure quite different from what the financial world had seen before"–especially his gregariousness and "open door to the press."[22] The reason for that sociability, however, was that he depended on public attention for his self-esteem. The outside world's response to him was a vital mirror. Transformed into his shrine at his death, his office was filled with souvenirs–an inspirational painting that Carnegie gave him, celebrity photos and letters of commendation–that asserted his status and worth. He kept volumes of press clippings and mementos that recorded his successes and scandals. He sought recognition as much as power and heatedly demanded his due when challenged.[23] All were factors in the subject matter of Gérôme's bronzes.

Schwab also admitted that he could not take orders. His Bethlehem Steel venture sprang, he said, from his desire to be a corporate "czar."[24] Used to the freedom he had enjoyed under Carnegie, [25] he balked when restrained. As he repeatedly said later, his honeymoon as president of U.S. Steel ended within months (mid-1901), when the executive committee thwarted his efforts to manage the company without consulting the board of directors and two operating committees.[26] The difference in management styles and struggle for control, he argued, were the primary reasons he retired as president two years later (1903) to develop the operation in Bethlehem.

However, an especially bitter public assault intensified this critical moment. Already notorious for aggressive business tactics, Schwab suffered, beginning in 1901, from a press attack that further jeopardized his reputation. Amid turmoil over his newest business dealings, the press also accused him of playing high-stakes roulette in Monte Carlo while on an extended European vacation during his first months in power, which he denied without convincing anyone who knew him. He had played high-stakes poker while at Carnegie (secretly, to prevent reprisals from his puritanical boss/mentor) and became a masterful whist and bridge player. Vilifying him as a reckless gambler, the press questioned Schwab's reliability with corporate funds. Upset at his seemingly unresolvable conflict with U.S. Steel and at the furor over a "vice" that alienated his most precious ally (Carnegie), the hypersensitive Schwab became ill and fled to Europe for the better part of two years, seeking cures and sympathy as U.S. Steel executives debated his already precarious fortunes there.[27] Though handled diplomatically in public (the official reason for his resignation was his health), he left U.S. Steel haunted by criticism.

Moving to Bethlehem to focus on his new enterprise, Schwab also plunged into a community wary of outsiders, especially controversial moguls. He immediately sought to fit in and make a contribution. He purchased the home of an eminent citizen who had just died, G. B. Linderman (704 Delaware Avenue, Fountain Hill), and took on civic work. Years later, he was credited with helping to mold the present Bethlehem from several of its boroughs and to build the Hill-to-Hill bridge that joined its two principal sections. World War I disrupted his program to create a huge community center on the property that had previously belonged to John Fritz, a venerated local steelmaster who was vital to the growth of what became Steel.[28] Schwab is best known for creating Steel's employee activity groups (the Bethlehem Steel Company Band, for instance, and various sports teams) whose skill gained fame well beyond the Lehigh Valley. However, he also became prominent in area cultural spheres, as a trustee of Lehigh University and especially as a music patron. He brought international figures to perform with the Lehigh Valley Symphony Orchestra and was, from its earliest years, chairman of the board and main guarantor of the Bach Choir. He arranged for the Choir to perform in New York in addition to its annual festivals, as he was fond of mentioning in his later years.[29] Rana was elected to the board of trustees of the Free Library of the Bethlehems in the midst of a strike and another campaign to unionize Steel in early 1910.[30] One local obituary noted that Schwab's community work made him Bethlehem's "first citizen" from the beginning.[31] But in the process, he also helped a region going through profound expansion and change, thanks to his programs at Steel, to achieve a new sense of community and local pride.

The arts were an important feature of Schwab's vision of professional achievement. Succeeding in business, he maintained, owed much to an imagination that grew with exercise in the arts.[32] His own passion, shared with his family, was for music. Since his youth, he had stood out for his singing voice and prowess at the keyboard—he had caught Carnegie's musical ear as well as professional eye—and even taught music early in his career for extra money.[33] He felt one's personal artistic interests need not be widely known. Many people, he said, were surprised to learn of Henry Clay Frick's stunning art collection in New York, which he made available as a museum there (still one of the finest in America) when he died in 1919. Thought to be merely "an extremely brainy man of business...," Frick would never have "attained the great place in industry which he did attain" if he were not "intensely interested in the beautiful and the artistic."[34]

Schwab himself, nonetheless, drew public attention for his collections of painting and sculpture. Completed in 1920, his country estate "Immergrün" garnered praise for its extraordinary gardens, featuring an important selection of contemporary American sculpture by John Gregory, Henri Crenier, and especially Paul Manship, which earned him public and critical praise for supporting the American avant-garde as a modern Medici.[35] However, the press most often mentioned the quality of his New York collection.

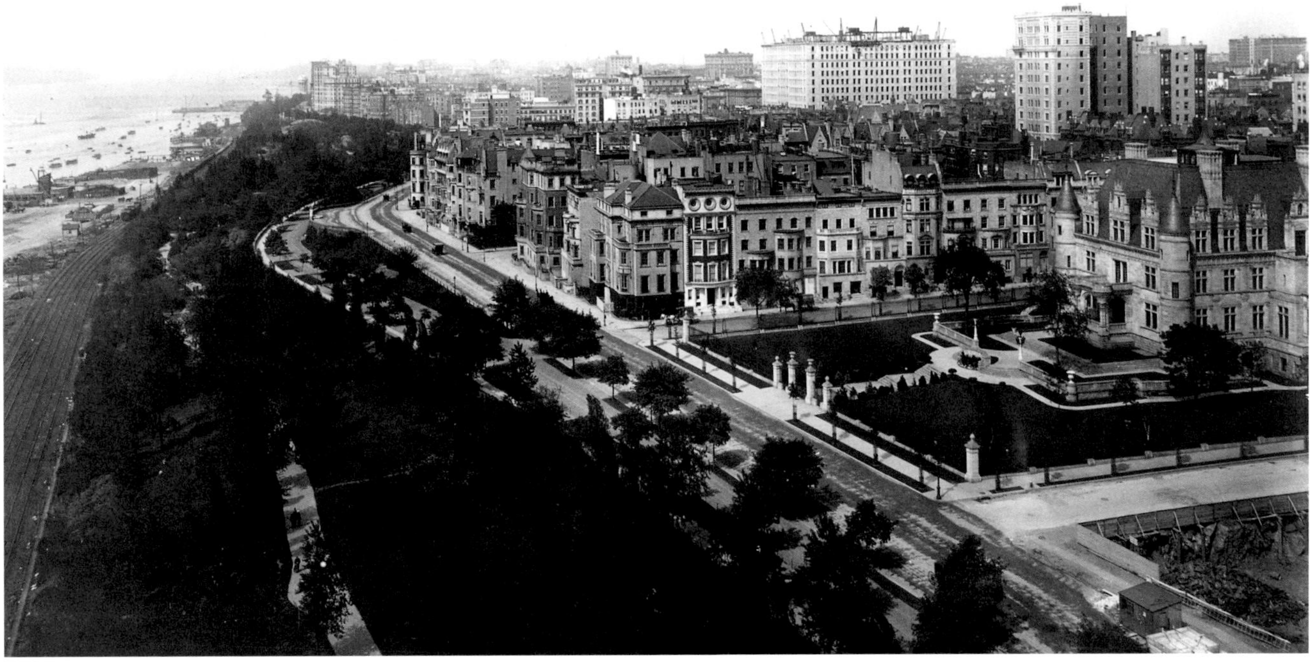

FIG. 4.

Aerial view of "Riverside" by 1920, Museum of the City of New York

Schwab seems to have collected paintings with varying degrees of ambition over time. Though much remains to be learned about the subject, he appears to have begun buying while in Pittsburgh, at which time he bought six paintings from Knoedler & Company in New York, some of which were kept until he died.[36] By 1904 he aimed higher. He bought a "Rembrandt" in Paris that was revealed as a recent fake that same year.[37] In 1906 he rallied and acquired a well-established Old Master work, Titian's portrait of Pietro Bembo, that the Samuel H. Kress Foundation later purchased at one of Schwab's posthumous sales and gave to the new National Gallery of Art in Washington, D.C.[38] He also pursued nineteenth-century avant-garde works. In 1917 he bought J. M. W. Turner's late *Rockets and Blue Lights* (c. 1840, Sterling and Francine Clark Art Institute, Williamstown).[39] Schwab made many such decisions jointly with Rana, who voiced her displeasure when excluded, as in the case of the Turner, which they kept nonetheless.[40] Shortly thereafter they apparently downshifted, using longtime associate Henry Frick as the model they chose not to emulate. On September 24, 1917, Schwab told Roland Knoedler to look for less ambitious paintings than Frick's: "I want good typical examples, of course, but not such extravagant pictures as Mr. Frick's, which pictures must be left to the public and must be of the highest type of excellence, but at the same time I desire good pictures." Though there is no record of any purchases after 1920 at Knoedler, Schwab asked the dealer to hold a painting in August 1923, though months before he said that he must stop buying because of his debts.[41]

Most of his considerable fortune disappeared after the Depression, thanks in part to bad investments outside his business and in part to his insistence on maintaining his expensive properties and supporting his native region.[42] At his death, Schwab was so deeply in debt that it took his executors years of careful management to settle his estate.

HISTORY OF THE BRONZE FIGURES

Gérôme's pair of statues were important features of the house Schwab built in New York City when he became president of U.S. Steel. Begun in 1902 and completed by 1907, the elaborate residence (Fig. 4) made a powerful and much-publicized statement.[43] Where it did not break with tradition, it simply went beyond all existing standards. It was bigger and flashier than anything built in New York up to that date, especially by established corporate executives. Outstripping the Carnegies' restrained new mansion and gardens (Fifth Avenue at East 91st Street, now the Cooper-Hewitt National Design Museum), the Schwabs' estate, formally called "On the Hudson" but dubbed "Riverside" for its location, was to be a massive château and garden that occupied a full block between Riverside Drive, West End Avenue, 73rd and 74th Streets, the only New York house to occupy such an expanse. The Schwabs had rejected the recognized neighborhood of the élite on the East Side for Frederick Law Olmsted's Riverside Park and Drive then under construction on the upper West Side, with their highly public promenades and extraordinary view of the Hudson River.[44] The design gave a "royal" French historical form (Renaissance façade elements from royal châteaux on the Loire and interiors that ranged into the eighteenth century) to a house intended to be, according to Schwab, "comfortable. Everything...subordinated to convenience, simplicity and wholesomeness in the living arrangements."[45] Much was made of the advanced technology. The house boasted its own electrical plant to operate an innovative ventilation system with individual thermostats in each room, three elevators, refrigerated cellars, and of course, a steel frame used before then only in commercial or industrial buildings, a "wonderful combination of art and science."[46] The Schwabs chose European architects and artists to design and execute the house and gardens. The architect-decorator, a specialist in historical exteriors and interiors, was Frenchman Maurice Hébert, the cousin of a power at the Institute of France and the Academy of Fine Arts, painter Ernest Hébert.[47] The main decoration of the Grand Hall was a lunette mural by Ernest Hébert's student, Albert Mantelet, with a ceiling and a projected garden fountain by Spaniards José Villegas (director of the National Museum of Painting and Sculpture in Madrid) and Fernando Miranda, with paintings and tapestries in various rooms (the parlor and dining room, for instance), designed by German-born and -trained muralist Conrad Arthur Thomas.[48]

Sculpture played a prominent role in Hébert's design. The interior teemed with figurative architectural ornament, and there were plans for freestanding sculpture that was apparently not executed.[49] For all its eclecticism—the Schwabs chose each room's style—one critic identified a subject that unified the entire project, inside and out: "the dignity of toil and the progress of American history," presented mainly through allegory.[50] "Riverside" thus celebrated a vital element of the American moral core, its work ethic.[51] The property's public spaces provided relevant themes: Miranda's fountain conceived for the "front," the Riverside Drive terrace (not executed), represented the *Fruits of Toil*; Mantelet's mural, *The Renaissance Distributing Laurels* (to builders and architects); Villegas's dining room ceiling, *American Prosperity Sending Her Industry, Agriculture and Abundance All Over the World.*[52]

Commissioned the year construction began on the house (1902), Gérôme's monumental bronzes added a specific industry (American steelmaking) to the site's generic theme—apparently the one artistic reference to Schwab's signature industry there—and a "modern" realist language that appeared nowhere else in Hébert's published designs and models.[53] The basic history of the figures is clear. Hébert approached Gérôme with the commission around the summer of 1902 and the Schwabs approved the preliminary maquettes on a visit to Gérôme's Paris studio that fall.[54] Now untraced, the plasters illustrated here were probably casts of some of the first clay sketches for the project (Fig. 5). The figures' basic concept remained relatively constant. For the attributes established at the outset, Gérôme used photographs of steelmaking equipment sent by the architect.[55] As the realist that Gérôme purported to be, his model for the *Metallurgical Worker* was a living American puddler in working garb whom Schwab had sent from Pittsburgh (Fig. 6).[56] Changing the figure's stance slightly from that in the sketch, the sculptor positioned his model so that he faced forward, rather than to his left as before, spread his feet, and further extended his right arm for the dynamic, yet balanced pose of the

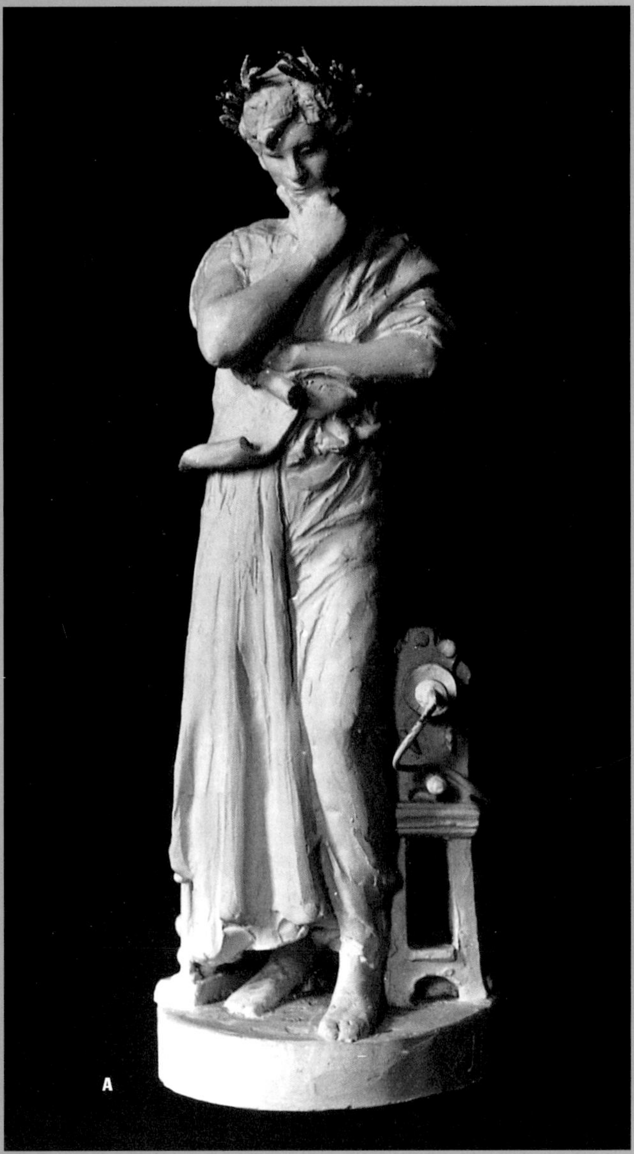

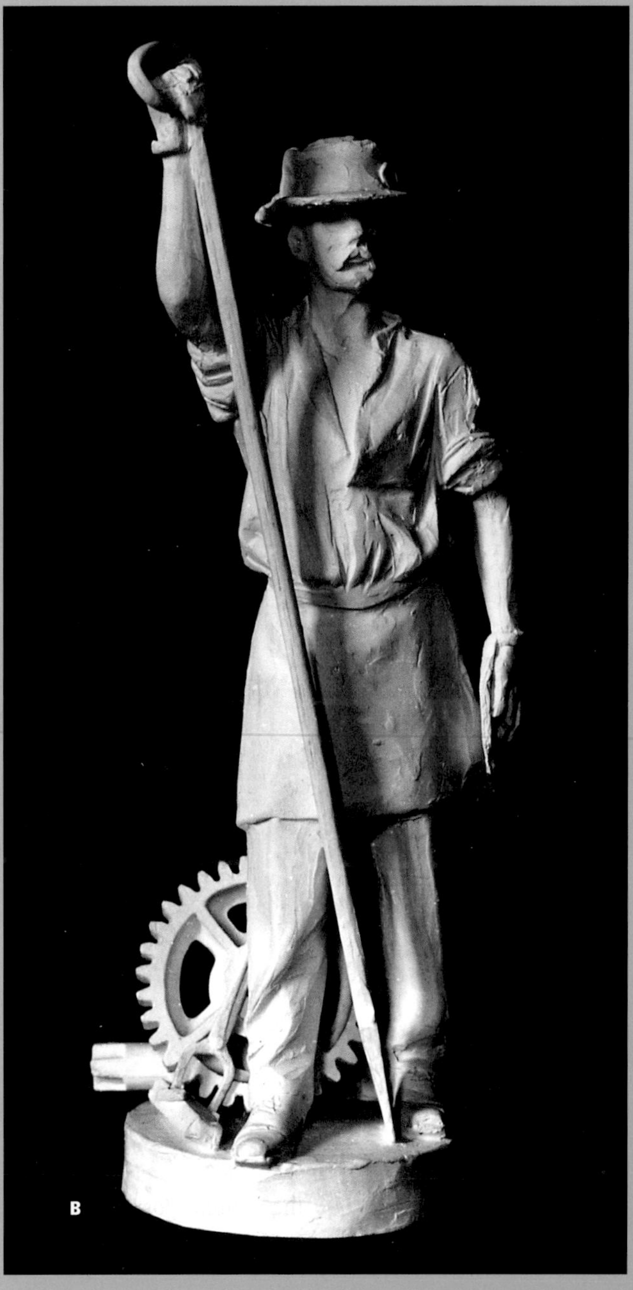

FIG. 5.

"Models for Art Gallery Group (by Jean Léon Gérôme)." Maurice Hébert, *"On the Hudson": Residence of Chas. M. Schwab, Esq.* (New York: privately published, 1905), vol. 2, leaf 81, Bethlehem Steel Collection, National Canal Museum, Easton, PA.

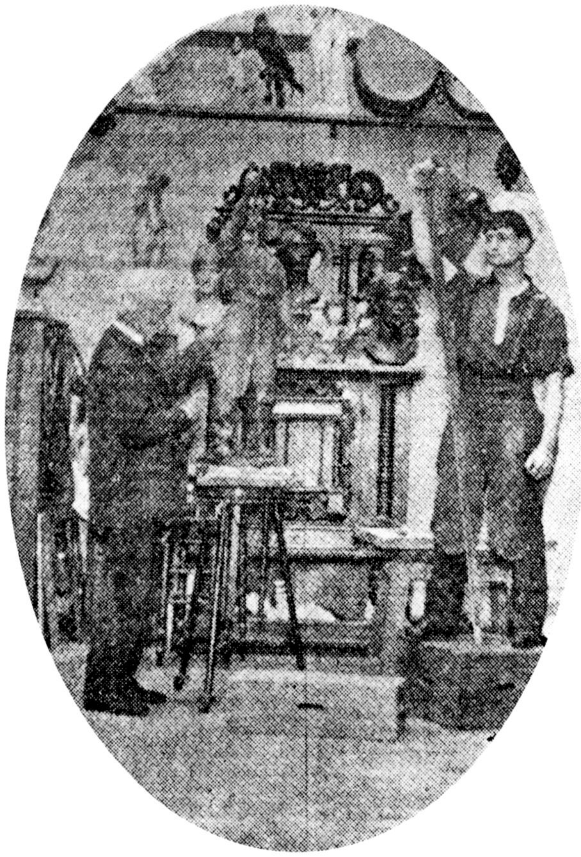

FIG. 6.

"Gerome Working with Model," C.L.B., "For Mr. Schwab's House," *New York Tribune* (Feb. 1, 1903), illustrated supplement: 5; in Hébert, *"On the Hudson,"* leaf 6. Courtesy, The Winterthur Library: Printed Book and Periodical Collection

finished work. However, the sculptor rejected the truth of the model for other features presented in the sketch: the hat and mustache, greater maturity, and sharp features of the worker's face. For *Scientific Metallurgy,* Gérôme directed the overall energy to the face: he reduced the body torsion, rolled up the scroll, and gave the eyes their unnerving focus (Fig. 1).[57] During the summer of 1903, Gérôme sent Hébert photographs of the definitive models, now enlarged to full scale and headed for the foundry.[58] The bronzes were apparently finished by the following December.[59] Despite the artist's death only weeks later, the bronze *Metallurgical Worker* appeared, as he had planned, in the next Paris Salon, a prestigious juried annual art exhibition that drew international artists and visitors. The pair was described as in place in an article on the Schwabs' unfinished house in 1905.[60] The total price of the project is unknown but it was undoubtedly high, given the great size of the two figures: Hébert paid installments of 25,000 French francs, enormous sums then as now.[61]

Other Gérôme sculpture was slated for "Riverside" as well, though we know little about it. It seems that Rana wanted some smaller pieces she had seen in addition to the monumental pair. The sculptor asked the architect for payment for the "marble figurine for Mme. Schwab."[62] Later Gérôme told Hébert that he was holding "two statuettes" for the house.[63] These tantalizing statements are all we have. What were these sculptures? Were they actually sent? If so, what happened to them? There is no sign of them in Schwab's papers or posthumous sales.[64]

It is tempting to speculate whether one of them was a marble that Ackerman called *Loie Fuller: Nude Unveiling Herself,* but which Hébert entitled *The Bather* in the Schwab Albums (Fig. 7).[65] The only version of that model known today measures thirty-nine inches. On the other hand, Schwab's may instead be a hitherto unknown example. Details in the Schwab Album photograph suggest it was larger, perhaps even life size.[66] Other evidence raises the further possibility that it was intended for Charles Schwab himself, rather than his wife. Hébert's elevation drawing of the east wall of his bathroom shows two niches with large classicizing female figures on pedestals, one a nude bather "testing" the water and the other dressed, lifting off (or putting on) clothing.[67]

The choice of Gérôme for the steelmaking subject, however, seems out of character with the overall strategy of the project. Unlike other artists involved, he was famous and well into his twilight years by 1902.[68] A student of painters Paul Delaroche and Charles Gleyre in the 1840s, he launched his reputation in the Paris Salon of 1847 with *The Cock Fight* (Musée d'Orsay, Paris), a large classical genre painting (image of everyday life in ancient Greece) executed in high-keyed, luminous color, clear contours, and smooth, invisible brushwork. It was neither academic nor subversively "vulgar," like the harshly realist works about modern life among the working classes by Gustave Courbet, for instance. Instead it offered a moderate approach that certain quarters of the public and state could accept. The painting was awarded a medal and bought by the government; it also introduced a style that came to be called "Neo Greek" or "Pompeian." In the 1850s, using the same color and brushwork, Gérôme expanded into so-called Orientalist subjects of modern life in Africa and the Middle East, inspired by his travels there, then delved into historical themes and dramatic deaths (Roman gladiators, the assassination of Caesar, and the execution of Napoleonic Marshal Ney). The artist was showered with honors, receiving the highest awards, and was made a member of the élite Institute of France and professor at its art school, the Academy of Fine Arts. He was less implacably hostile to the avant-garde than is currently thought, however. Though he opposed the gift of Gustave Caillebotte's Impressionist paintings to the French nation in 1894, he publicly praised the work of Auguste Rodin and Claude Monet.[69]

Elegant, urbane, and rich (thanks to his wife, the daughter of his prosperous dealer), Gérôme was in demand socially as well. He hunted with the Rothschilds and was a regular at the country home of France's sovereigns, the Emperor Napoleon III and his wife, the Empress Eugénie, and at the gatherings of the Emperor's cousin, a patron of the arts, Princess Mathilde. Yet he also had close friends among the avant-garde, including Edgar Degas and Monet. Secure in his career and restless intellectually, in the late 1870s he returned to sculpture—he had reportedly carved ivory as a child in the studio of his father, a goldsmith, and modeled two small gladiators early in his career—and turned to sculptor friends to learn large-scale techniques.[70] Animated by his signature scientific realism, anatomical suppleness, and sharp aesthetic sense, his three-dimensional work proved as successful as his paintings. The prolific artist produced many lively figures of ancient and modern subjects (including public monuments), many of which he marketed as multiples, primarily in bronze. He also triggered considerable debate among the critics about realism and color in sculpture, with riveting works in colored stone and finishes. He took yet another creative turn with a dramatic allegorical monument at Waterloo, commemorating Napoleon's defeat there. He was engrossed in work when he suddenly died at eighty in January 1904.

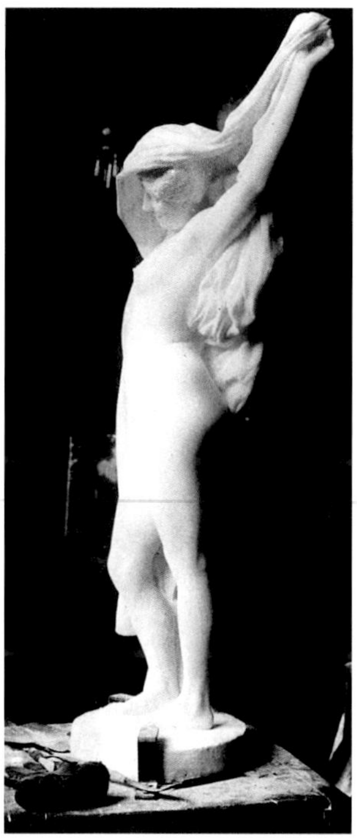

FIG. 7.

Gérôme, *Bather*, c. 1903, in Hébert, *"On the Hudson,"* folio, leaf 82, Bethlehem Steel Collection, National Canal Museum, Easton, PA

For all the luster of his name and his versatility, Gérôme's art seems worlds apart from Hébert's concept for Schwab's mansion. The eerie lifelikeness of his sculpture had little to do with the ideal Renaissance-inspired work of the other artists employed at "Riverside." The given subject was far removed from Gérôme's repertory of themes. He was not among the "contemporary" practitioners of sculpture that a French authority touted at the time to American architects and patrons.[71] And, most of all, why not one of the senior European sculptors (Jules Dalou and Constantin Meunier) who had built their reputations on tributes to modern labor? Whatever the reason, Schwab may have been swayed by an unrelated passion that he shared with the architect and artist. All revered Napoleon; Hébert knew Gérôme in part because he too socialized with his descendant Princess Mathilde.[72] Timing favored that passion: When the Schwab project began, Gérôme was just finishing his Napoleon monument for Waterloo.[73]

The giant bronzes for Schwab were conceived for a specific site: the Louis XIV-style art gallery on the second story of the northeast wing, facing West End Avenue.[74] That is where they actually stood by 1905—possible, given their great size and weight, only because of the house's grand scale, elevators, and floors supported by structural steel. In a gallery officially devoted to paintings (Fig. 8), the monumental figures stood out in every respect. One admirer remarked: "The cupids and roses, looped ribbons and torches and all the graceful symbolry of the past are struck into insignificance by these massive representations of the masterful present. Towering up on their marble pedestals in the art gallery, they dwarf all else."[75] There, standing in the center of the gallery (like the sculpture in Hébert's watercolor), they could be examined up close and fully in the round, revealing the excellent and subtle handling of their surfaces. The art gallery and its content also emphasized the figures' identity as high art, surrounded, as they gradually were, by a Hals genre piece purchased in 1902, the Titian bought in 1906, eighteenth-century portraits, and some distinguished progressive works of the nineteenth century, especially the Turner.

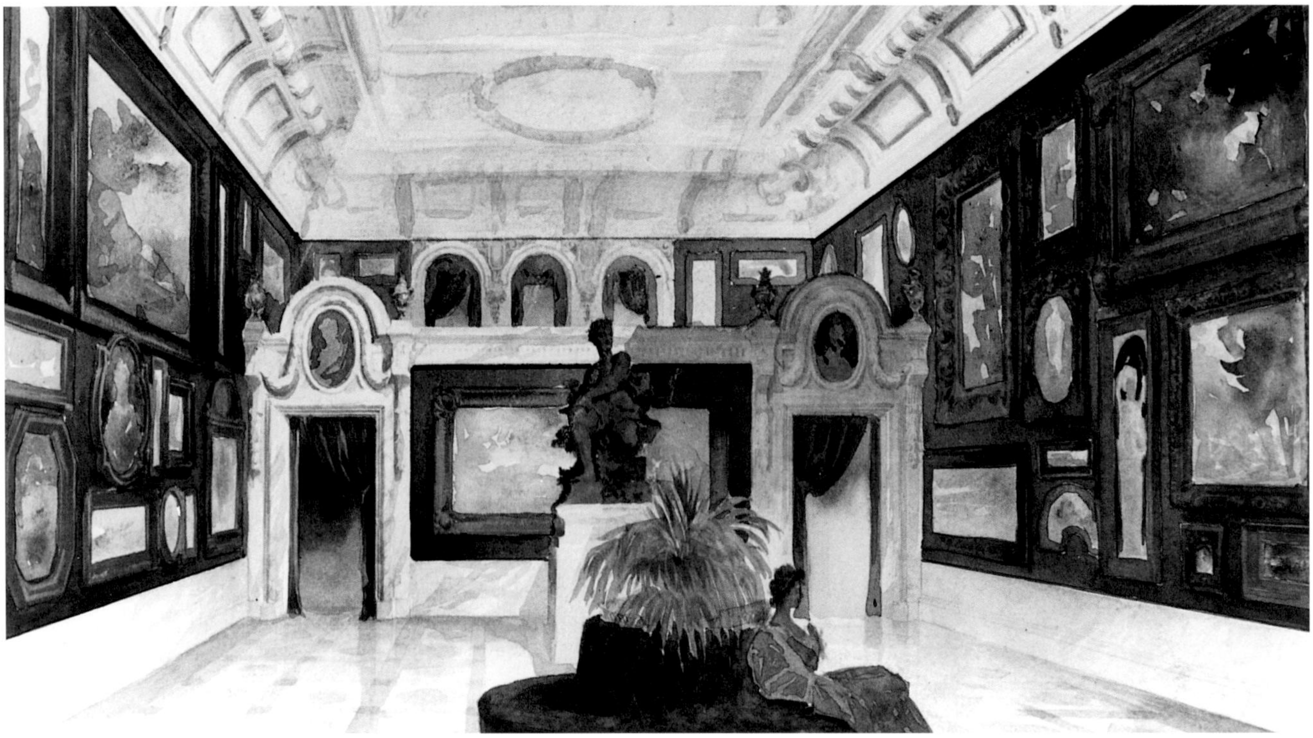

FIG. 8.

Hébert, "Picture Gallery–Perspective View," in Hébert, *"On the Hudson,"* leaf 21, 1983.148.460, Bethlehem Steel Collection, National Canal Museum, Easton, PA

At an unknown date,[76] the two bronzes were moved outside, to the west terrace facing Riverside Drive, where they stood as the house awaited demolition in 1947. Visible to all who strolled in the park or drove by, the figures gained a public presence that they had not had indoors, blurring the threshold between private and public realms. The bronzes visible in these open spaces became quasi-public monuments to this "steel king," as Europeans called Schwab,[77] in his private domain within the burgeoning city.

Moved to Bethlehem Steel because of their link to Schwab, the bronzes awaited installation there until in 1982 the company instead gave them to the Allentown Art Museum, where their placement and new historical moment subtly shifted their meaning. Facing the city beyond, they became exceptionally fine "museum" pieces that paid public homage to a man and an industry that had strongly shaped the Lehigh Valley's modern identity and was now in its twilight.

ORIGINAL MEANING OF THE BRONZES

Schwab had a special job in mind for these figures and, though Gérôme may have suggested ways to convey it in sculpture, the steel magnate established their concept himself.[78] He wanted to represent the disciplines that he felt brought American steelmaking to its glory, disciplines that were closely identified with him. His bronzes were thus monumental tributes to himself as a steel man. According to Gérôme, Schwab ordered figures to "symbolize in the most striking manner [his] life and destiny."[79] A journalist who interviewed Schwab in the nearly finished house wrote that their subjects were deeply emotional for his host: "These huge bronzes represent the sentimental side of the master of the house, his struggles and his victories."[80] When this comment was made (1905), Schwab was in his first year as chairman of Bethlehem Steel. However, the year he commissioned the figures (1902) he had run to Europe from a much-publicized professional crisis: his struggle for power at U.S. Steel and for his reputation in the international business world. In this light, Gérôme's figures demanded recognition for Schwab's accomplishments in the face of strong criticism and thus gained a special urgency, even poignancy.

The two bronzes were allegories of Schwab's often-repeated business principles. Together, as *Metallurgical Science* and *Metallurgical Worker,* they represented the two forms of metallurgical knowledge or expertise demanded by successful modern steelmaking. The male worker embodied traditional knowledge gained on the job with hands-on experience. The female conveyed "modern" knowledge provided by theoretical science, which added new methods of steelmaking to the old and, together, placed America at the forefront of the industry.[81]

Schwab had witnessed the strengths that each approach brought to the industry during his first years there. Traditional steelmaking involved sensory analysis: judging, at extremely high heat, the state of molten metal by its feel, smell, and sight. The newer Bessemer and open-hearth methods made the process especially volatile and dangerous, capable of triggering a furnace explosion or producing flawed steel.[82] The financial consequences were as serious as the safety hazards. The Pennsylvania Railroad once rejected a shipment of steel made at Braddock, while Schwab was in the engineering department there, because it did not meet the specifications in quality and alloy required by the Railroad's president, who had begun to study steel chemistry because so much of his equipment (especially steel rails) was made of the metal. Schwab's response was to search for possible technical controls over manufacture through chemical experiments at home. Though primitive, they convinced his superiors of the importance of metallurgical study to steelmaking and of the ingenuity of this young employee. By the time he became president of U.S. Steel, Schwab had made the story of his chemistry experiments in his wife's sewing room part of his personal success story, as a sign of the power of initiative and the desire to solve problems through learning.[83]

His support for metallurgical science was also an endorsement of self-education. With only some high-school-level certificates and no formal technical or college training, Schwab regularly argued that a college education did not guarantee success; only personal initiative and the pursuit of knowledge could prevail. Schwab thus set himself apart, as one of the self-starters, from the executives who went to Ivy League universities.[84] *Metallurgical Science* further includes, at its base (Fig. 1), an image of the steam de-scaler (a machine used then to refine the surface of freshly-made steel ingots) to represent all steelmaking devices that Schwab supposedly improved or invented. These various references to his material contributions argued for the imaginative use of science. The scientific mind, he suggested, is not formulaic. The figure conveys the creative energy of such thinking in two forms. One is the abundant laurel that encircles her head as a garland and thrives at her feet as a bush. The plant identified with Apollo as patron of the arts, bay laurel is a classical symbol of divine inspiration or poetic genius. The second cue for creative energy is in the woman's fierce concentration. The familiar pose for meditation is made tense, channeling all focus into her huge, downward-turned head and sharply delineated pupils (cover illustration). Inner thought probes and analyzes the outer world. She radiates physical vitality as well, implying her direct involvement with labor. Her body is sturdy, with strong, muscular limbs, especially the formidable forearms that frame her head from our frontal viewpoint. She is thus a remarkable combination of physical and intellectual power.

Though the figure indirectly congratulates Schwab for these attributes, it openly praises the vitality of America's women. Gérôme's *Metallurgical Science* was to be "the symbol of American womanhood" through its physiognomic features: "a young American woman with high forehead, beaming, intelligent eyes, mouth firm and indomitable chin sharply cut and energetic."[85] Her male companion represents another ideal modern American type. Gérôme claimed it was his rare pleasure to try to "fix the traits and character of the real-living American workman with all his inherent national force and energy," using an authentic laborer who also met the aesthetic standards of great sculpture. He was "an American steel worker selected for his magnificent manly proportions and superb head and nervous [sensitive] neck, which are of the neoGrecian or American type."[86] As mentioned earlier, Gérôme altered the physical reality to suit his vision, apparently without objection from Schwab.

This was, moreover, an image of the American worker's dignity and pleasure in famously brutal work: "M. Gérôme has imprinted this laborer's face with a joy and energy that suggest with what heart the most difficult work is carried out in the United States and what hopes sustain him."[87] Such a view gives a glowing picture not only of a grueling industrial job but of an American labor force that at the time was still in profound turmoil, with strikes, conflicts with management, and efforts to expand unions as the industry grew and changed, thanks to new processes and the European immigrants hired for unskilled jobs.[88] Here too is a tacit tribute to Schwab, whose rapport with, and influence over, workers was legendary and often crucial in the many labor confrontations he had faced by 1902—notably the various unsuccessful campaigns to unionize his workforce.[89]

Gérôme represented the American worker, however, by an especially famous and masterful craftsman among them, the puddler, shown "vigorously handling a sort of lever [the paddle-shaped rabble]...with which he stirs molten metal."[90] He does not make steel itself; rather, he creates the wrought iron that provides the raw material for crucible steel, the highest grade of this "imperial metal" reserved for edged and metalworking tools and crucible steel bars, saw plates, and sheets.[91] Gérôme suggested the puddler's pivotal role in the process with the ingots to the rear, and originally a large gear and turning shaft (Fig. 5b), included perhaps to invoke the machinery used for working crucible steel after casting.[92] Unlike the scientists who made open-hearth and Bessemer steel viable, the puddler brought what were considered ancient human strengths to modern industry.

Gérôme's monumental figure conveys the puddler's most important qualities. Intelligent and keenly alert, he was modern industry's counterpart to the classical athlete whose physical attributes were defining: strength and endurance, coordination, and judgment based on senses trained on the job. James Davis, a master puddler of the 1890s who went on to become secretary of labor under Warren Harding (1921–25), glorified this craftsman's physicality with extravagant rhetoric. He proclaimed that: "mental knowledge is not enough; it requires great muscular skill like that of the heavyweight wrestler, besides great physical endurance to withstand the terrific heat. The worker's body is in perfect physical shape and the work does not injure him but only exhilarates him."[93] Davis was proud of the physique with which this craft provided him as he matured: "I went in [to the rolling mills] a stripling and grew into manhood with muscled arms big as a bookkeeper's legs."[94] He gave a detailed account of puddling in the 1880s and 1890s at the rolling mills of Sharon, Pennsylvania, in order to prove the intelligence and many skills involved, including the proportional judgment of "the artist and the sculptor."[95] Workers, he explained, put six hundred pounds of pig iron in the intensely burning coal furnace, which they alternately stoked and stirred with a long iron rabble that weighed about twenty-five pounds. After about thirty minutes, the puddler "thickened the heat up" by adding roll scale (a form of iron oxide promoting chemical reaction) to the bath, stirring it in with the rabble, protecting his bare hands from the dangerously hot metal rod by his thick hand-rags and boiling out impurities in his "puddle." As the pig iron melted, he reduced the heat to "oxidize the phosphorus and sulfur ahead of the carbon," gauging the reaction by the "light flames breaking through the lake of molten slag in my furnace." Stirring constantly, the puddler poured off slag as the molten metal boiled. When it became a stringy sponge of pure iron, he rolled it into three balls or blooms that weighed about two hundred pounds each and rushed them, clamped in long-handled tongs, to the squeezing machine to release more slag. The puddled iron was then ready for remelting to make crucible steel, a process undertaken by another immensely strong specialist.

Perhaps chosen for his prestige, as the most skilled and highly paid worker in the industry, the puddler was nonetheless in a minority among steelworkers by the early twentieth century. Because the manufacture of crucible steel of which he was a part was labor intensive, costly, and thus geared to highly select uses, it represented only a small part of an industry dominated increasingly by the soft-steel products made by the open-hearth and Bessemer processes and by the unskilled workers involved. Considered the steel industry's élite by some, the puddler and his craft were viewed by others as archaic, signs of a company unwilling to embrace the rapid, if risky, technological advances in the modern steel industry.[96] By the early twentieth century, some argued that the art of puddling was under threat. Despite the simplicity of the process and crudity of the furnace, the sheer skill demanded of a good puddler was the kiss of death to this craft, making it increasingly difficult to attract intelligent men to such a demanding profession when others involved less effort and risk—and ultimately paid better.[97] Schwab may have capitalized on the puddler's almost mythic aura when he referred to himself in speeches as just a "steel puddler from Pittsburgh."[98] Had the puddler joined American heroes like the frontiersman and the cowboy? The pantheon of American icons gains diversity with industrial figures like the puddler, and later the riveter (Fig. 3) who rose to fame in the shipbuilding push of World War I, thanks in part to Schwab.

Thus, Gérôme's two figures represented not only the equal status of two types of knowledge (theory and practice or intellectual and sensory) but also a pairing of the modern and the traditional. By such means Schwab conveyed his complex vision of himself in relation to an industry—and to modern life—in profound flux. He may also have asserted his commitment to puddled steel for the future. He later advertised puddled bar and tool steel among the diverse offerings at South Bethlehem.[99] Finally, even the medium in which the pair was executed subtly relates to Schwab's subject. Made of cast bronze, an art form that deploys molten metal and the tooling and coloring of the cooled substance, the figures celebrate metalmaking through the symbolic language of material and technique.

THE BRONZES AND THE ART WORLD

These two statues are unique in Gérôme's work for their subject but have features that are fundamental to his art. Some of those qualities are stylistic: their supple vitality and naturalistic truth, thanks to the artist's strong emphasis on direct observation and scientific learning and the aesthetic controls that provide both their drama and subtlety. As allegories, the pendants reveal Gérôme's probing of symbolic forms in his last years. For Schwab, the sculptor used his signature vehicle, the human figure. Nonetheless, Gérôme chose to make a public statement by pairing the *Metallurgical Worker* with another figure, his *Corinth,* for exhibition at the Salon.[100] Despite their similar realism, the works are striking in their differences. The dressed worker, symbol of modern industrial life, well over life size and of dark patinated bronze, contrasts with the nude female, symbol of an ancient Greek city famed for its courtesans, small in scale (a twenty-six-inch figure on a nine-inch column), and precious in execution (polychromed marble, semiprecious stones, and gilded bronze; private collection, Paris).[101] Together they demonstrated Gérôme's remarkable versatility and mastery of his craft, as well as his commitment to pursuing new directions despite his success and age.

Alone among his works, Gérôme's *Metallurgical Worker* can be linked to avant-garde images of modern industrial labor that proliferated in Europe and America, often to wide acclaim, despite the tensions caused by industrial growth. Encouraged by the Enlightenment's endorsement of physical labor and industrial activity as socially useful, images of classical gods, heroes, and agricultural workers went modern as artisans and factory laborers in various progressive styles by the 1880s. There were paintings in the luminous colors and dynamic rhythms of "pointillist" Neo-Impressionism (Maximilien Luce's *Iron Foundry,* 1889, Rijksmuseum

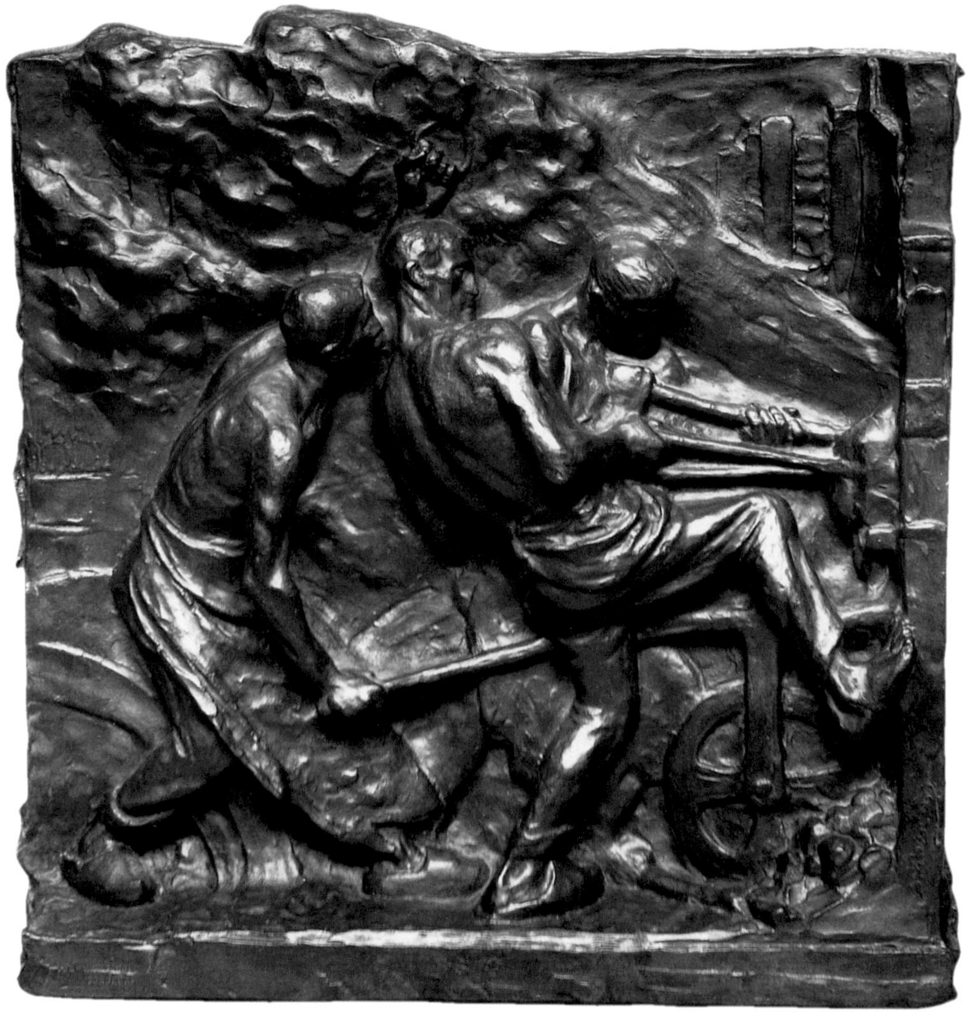

FIG. 9.

Constantin Meunier, *Puddlers at the Furnace*, c. 1893, Musée d'Orsay, Paris. Copyright Réunion des Musées Nationaux/Art Resources, New York

Kroller-Müller, Otterlo) as well as the gritty earth tones of realism. Progressive sculpture—monumental and intimate—followed suit with images of labor and workers, in reliefs and figures by Vincenzo Vela in Italy and, in France, Jules Dalou, who died the year Schwab commissioned his steelworking subjects.[102] Gérôme's subject matter, however, draws him especially close to an artist whose paintings and sculpture of workers profoundly affected Europe and America, the Belgian Constantin Meunier (1831–1905).[103] He was especially admired in France, thanks in part to his puddlers: The government purchased a bronze relief of the subject for the nation's art museums in 1893 (Fig. 9). Meunier elevated prosaic reality to a poetic modern symbol through an alliance of delicately rippling "modern" surfaces, traditional sculptural principles, and modern subject matter. However, he also dignified labor as "lived" experience, as he called it, producing expressive images that dismissed the edgy polemics surrounding labor to emphasize shared human experience.[104] Gérôme's athlete-worker likewise transformed a modern social ideal into a modern sculptural monument, but in far bolder, more virile form.

As an émigré to the United States, *Metallurgical Worker* fit especially well among American monuments to labor for its "masculine" qualities. With its classical mate, it distilled essential features of the first among them, such as Johannes Gelert's celebrated *Police Monument* (1889, bronze, Police Academy, Chicago) and various sculptural projects that decorated buildings at the World's Columbian Exposition of 1893.[105] Gérôme's pair incorporated the two stylistic approaches to the subject of labor in the United States thus far, one traditional and one "modern," in order to propose the comprehensive merger of old and new, female and male. Though they look very different, the French figures share features with a special monument, the one known American memorial specifically to industrial labor, San Francisco's *Mechanics Fountain,* by California's first sculptor to achieve an international reputation, Douglas Tilden (Fig. 10). Both monuments celebrate vibrant physicality and industrial work through sturdy, heavily muscled bodies—Gérôme, with chaste realist clothing and implied activity rather than Tilden's ideal, erotically charged, straining athlete-workers. Both projects span private and public realms. Where Gérôme's bronzes paid homage to Schwab in the public spaces of his famous home, Tilden's fountain, located in a bustling commercial area, was James Mervyn Donahue's tribute to his father, industrialist Peter Donahue.[106] Both monuments honored the industry and its workers in ways that could touch a broad public despite the upheavals of modern life. Today, Schwab's earthy, confident, and eloquent figures make their presence felt, in the public spaces of the Lehigh Valley, as witnesses to profound change in their time and ours.

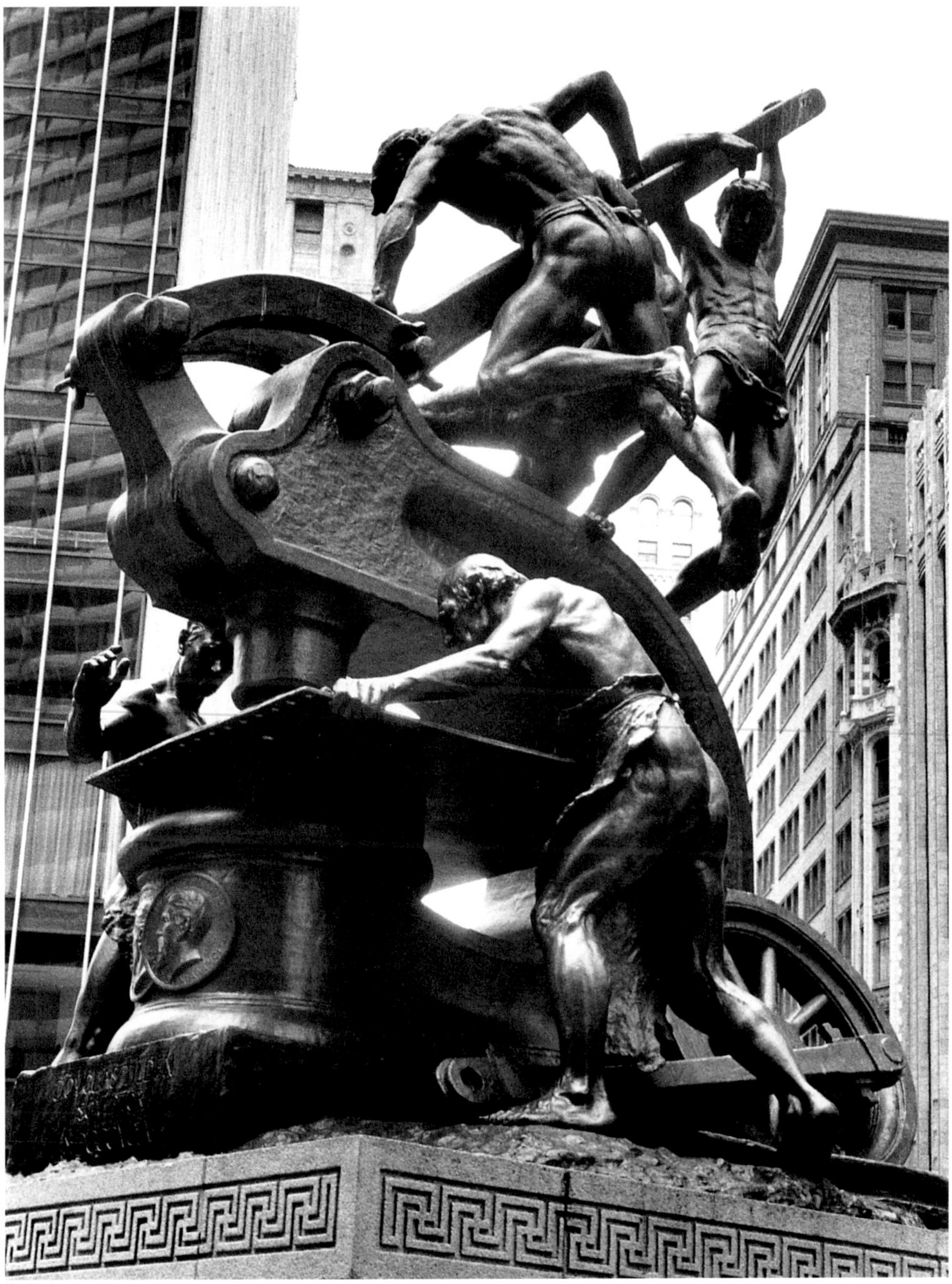

FIG. 10.

Douglas Tilden (1860-1935), *The Mechanics Fountain,* 1896-1901, Lower Market Street, San Francisco. Photo: Melissa Dabakis

NOTES

Unless otherwise noted, all translations are by the author.

1. Both suffered extensive losses owing to handling and corrosion from exposure to an acidic outdoor environment. *Metallurgical Worker* was apparently sandblasted in the 1940s, depriving its surfaces of their sensitive volumes and textures. Both bronzes were repatinated in 1989 to approximate their original handling, which apparently differed markedly from each other. The modulated light brown/blue-green patina on *Metallurgical Science* is partially visible through the new coat. Surviving traces of the original patina in protected areas of the *Metallurgical Worker* (at rear, thanks to storage face-up on a pallet) served as guides for its repatination.

Though titles vary throughout publications on these bronzes, those employed here derive from the ones most often used in the original correspondence and the catalogue entry for the *Worker* in the 1904 Salon catalogue.

2. As noted by the then-director of the Museum, Richard Gregg, in a letter to George R. Hall of Bethlehem Steel, dated May 19, 1982. See curatorial records, Allentown Art Museum.

3. His obituaries in New York newspapers develop this mythic quality of his life. See, for instance, "Charles M. Schwab Dies Here after Return, Ill, From Europe.....," *New York Herald Tribune*, 99, no. 133, 910 (Sept. 19, 1939), 1, 18.

4. Typical among the many examples is Charles Schwab, *Ten Commandments of Success: An Interview by B. C. Forbes with Charles M. Schwab* (Chicago: La Salle Extension University, 1920).

5. The most detailed biography to date is Robert Hessen, *Steel Titan: The Life of Charles M. Schwab* (New York: Oxford University, 1975). Others are currently in preparation.

6. See, for instance, "Schwab Dies at 77 in His Home Here...." *New York Times*, 89, no. 29, 823 (Sept. 19, 1939), 1, 25.

7. Eugene Grace, *Charles M. Schwab* (Bethlehem, PA: privately published, 1947), 5, was one of the few texts to accurately represent this facet. Grace was Schwab's protégé at Bethlehem Steel.

8. Hessen, 30.

9. For details, see Arundel Cotter, *Authentic History of the United States Steel Corporation* (New York: Moody Magazine and Book Company, 1916), 1–34.

10. "Charles Schwab, "President, United States Steel Corporation," *Bulletin of the AISA*, 35, no. 7 (April 10, 1901): 52.

11. *New York Times*, 25.

12. For a description of plans for his funeral, see ibid., 1.

13. Profiles of Schwab during his lifetime and long afterward emphasize this dimension. See, for example, "Thro' the Needle's Eye," *Black and White* (July 1918), Schwab United States Shipping Board albums; Bethlehem Steel Collection. These consist of a series of albums containing press clippings, correspondence, and photographs relating to his years as head of the United States Shipping Board. Among the most often-cited sources that illustrate this point is Grace.

14. *New York Times*, 25.

15. *New York Herald Tribune*, 18.

16. Ibid.

17. Schwab, unpaginated page opposite p. 2, 17.

18. Grace, 34, admitted Schwab was a spender who needed a "check-rein," Grace's unhappy job.

19. Schwab reportedly also had affairs with younger women throughout their marriage; see Hessen, 121–22, 292–93.

20. "Mr. Schwab's Mansion," *Real Estate Record and Builder's Guide* 73 (May 7, 1904): 1043. For Rana's alleged involvement in purchases of paintings, see below.

21. See the relevant clippings in the Schwab United States Shipping Board albums, Bethlehem Steel Collection.

22. Grace, 18.

23. For an example at the end of his career at Bethlehem Steel, when certain board members unsuccessfully sought to end Schwab's salary as nominal chairman, see Hessen, 296–97.

24. *New York Herald Tribune*, 18.

25. Grace, 13.

26. *New York Herald Tribune*, 18; Hessen, 125–28. For another assessment of Schwab at U.S. Steel, see Brian Apelt, *The Corporation: A Centennial Biography of United States Steel Corporation, 1901–2001* (Pittsburgh: Cathedral, 2000), 55–58.

27. For a discussion, see Hessen, 134–62. The subject of ongoing debate, Schwab's problems within U.S. Steel are most often discussed as centering on his philosophical differences with the corporation's top executive, Elbert Gary.

28. "Steel Chairman Had Been Failing for Past 5 Years," *Bethlehem Globe-Times*, 71, no. 22, 489 (Sept. 19, 1939): 10.

29. Ibid., 1, and Schwab, 16. For a brief recent discussion of his activities in Bethlehem, see "The Great Adventure of Charles M. Schwab," *Bethlehem of Pennsylvania: The Golden Years 1841–1920* (Bethlehem: Bethlehem Chamber of Commerce, 1976), 194–218.

30. "Free Library Trustees Elected," *Globe-South Bethlehem*, 16, 31 (Feb. 5, 1910): 8.

31. *Bethlehem Globe-Times*, 1.

32. Schwab, 16.

33. Grace, 8; Hessen, 7, 10–11, 19, 21, 285–87.

34. Schwab, 16–17.

35. For discussions of the gardens, see C. W. Leavitt [their designer], "The Gardens of Charles M. Schwab, Loretto, Pa.," *Country Life* (June 1920): 45–48; see also Robert Imlay, "The Gardens of Charles M. Schwab, Esq., Loretto Pa....," *Architectural Record*, 47, no. 5 (May 1920): 387–407. The sculpture was dispersed when the property was auctioned in 1942, with some pieces today in American public collections: Manship's *Hercules Upholding the World* is now at the Museum of Fine Arts, Houston, and his four terra-cotta vases, at Lincoln Center, Minneapolis. See John Manship, *Paul Manship* (New York: Abbeville, 80–83, plates 72–76.

36. According to the archives at Knoedler & Company, Schwab bought landscapes and figure paintings dating from the late nineteenth to the early twentieth centuries (present whereabouts unknown) between July 1898 and Dec. 1900. With the exception of works by William-Adolphe Bouguereau (*The Midday Meal*) and Félix Ziem (*Constantinople*), most are by painters who are little known today: R. W. van Boskerck, *Lily Pond, Wakefield, Connecticut*; Jan Van Chelminski, *The Heroes of 1812*; Théobald Chartran, portraits of Admiral and Mrs. Dewey; Alexei Harlamoff, *The Little Flower Girl*; and Cornelis Westerbeek, *Flock of Sheep on the Dunes*. Warm thanks to Edye Weissler, librarian/archivist of Knoedler & Company, New York, for her research on behalf of this project.

A posthumous sale also included a pastel by Frits Thaulow of the Carnegie Steel Works (*Carnegie Steel Plant at River Edge*, present whereabouts unknown) which may have been commissioned in Pittsburgh. See *The Eminent Collection of Oil Paintings by World Renowned Artists from the Collection of the Late Charles M. Schwab*, Tobias, Fischer & Co., Inc., Apr. 24, 1940, no. 16, repro.

37. Unidentified U.S. newspaper clipping for Feb. 25, 1904, Schwab General Scrapbooks, Bethlehem Steel Collection.

38. Knoedler & Company archives. For current information on the Titian painting, consult the relevant page on the National Gallery's website, www.nga.gov: the collection.

39. See Martin Butlin and Evelyn Joll, *The Paintings of J.M.W. Turner* (New Haven: Yale University, 1977), vol. 1, 216–17, cat. no. 387; vol. 2, pl. 361.

40. "The Turner which I recently bought, Madame does not like. I think the chief reason for not liking it, is because she was not consulted about its purchase"; Schwab to Roland Knoedler, Sept. 26, 1919; Knoedler & Company archives. I owe this selection of letters to Edye Weissler. All subsequent references to correspondence between Schwab and Knoedler are from this source.

41. Schwab to Knoedler, Feb. 8 and Aug. 24, 1923. Letters from 1921 mention his payment, with interest, on a note due—for what is not clear.

42. Grace, 42; Hessen, 299–300, 302–3.

43. The property was purchased around 1901 and ground was apparently broken in July 1902 ("Mr. Schwab's Great House Nearly Finished," *Real Estate Record and Building Guide* 77 [Feb. 7, 1906]: 288). Long delayed because of labor problems on the job, the house apparently received certain monumental elements like the Gérôme bronzes in 1905 (see below) and was reported as ready for the owner to take possession "in a few weeks" in early 1906 (ibid.). The house was published in its completed state in early 1907 ("Château Schwab," *Architectural Record*, 21, no. 8 [Feb. 1907]: 96–102).

44. "American Residence Series: The Residence for Charles M. Schwab...," *Architectural Record*, 12, no. 5 (Oct. 1902): 538–47; *Architectural Record*, 21, no. 8 (Feb. 1907): 96–102; Robert A.M. Stern et al., *New York 1900: Metropolitan Architecture and Urbanism 1890–1915* (New York: Rizzoli, 1983), 376–77; and Nathan

Silver, *Lost New York* (Boston: Houghton Mifflin, 2000), 126. David Brownlee directed me to the last two texts.

45. James Creelman, "First Photographs of Charles M. Schwab's $5,000,000 Palace in Riverside Drive," *New York World*, 45, no. 16,014 (June 25, 1905), magazine section: 6.

46. Ibid.

47. Hébert's specialty is listed on his letterhead; see Maurice Hébert, *"On the Hudson": Residence of Chas. M. Schwab, Esq.* (New York: privately published, 1905). During the Schwab project and afterward, Hébert listed three professional addresses in New York: in 1902, Flushing; in 1905, Astor Court, New York (possibly Astor House, a residential hotel on Broadway facing City Hall Park which predated the Astor Court built at West 89th and 90th in 1916); and in 1906–8, 20 West 34 St. The first and third are from James Ward, *Architects in Practice: New York City 1900–1940* (Union, NJ: J & D Associates, 1989), 34, a source brought to my attention by Sarah Bradford Landau. The Astor Court address appears in Hébert; the information on Astor House is from Robert A. M. Stern et al., *New York 1880....* (New York: Monacelli, 1999), 518; on Astor Court, *New York 1900*, 293. My thanks to Michele H. Bogart for help on this point. Hébert published a layman's primer on historical architectural styles "for Parlor Conversation" (*Brief History of Styles in Architecture and Decoration*) in various editions from 1895 to at least 1907. The only other architectural commission associated with Hébert to date is an unexecuted house, "Chateau Fresno", in Fresno, California, for the farm developer and vintner Theodore Kearney (1842–1906). Hébert's published designs for the project openly echo Schwab's house (*Chateau-Fresno....* [Paris and New York, 1907]). He was to complete a British architect's plan from the 1880s but Kearney's heirs, the University of California, did not move forward with the project, as Kearney had stipulated in his will. Hébert made headlines in Los Angeles in 1909, when he sued for an exorbitant unpaid fee. This information is from the docent manual provided by Michele Laurie, collections assistant, Kearney Mansion Museum, Fresno City and County Historical Society.

48. For illustrations of the models, see the Schwab Albums. For information on these now-obscure artists, see Hébert.

49. See the designs illustrated in the Schwab Albums.

50. Creelman, 7.

51. For a discussion of the American work ethic in the context of American art, see Melissa Dabakis, *Visualizing Labor in American Sculpture: Monuments, Manliness, and the Work Ethic, 1880–1935* (Cambridge: Cambridge University, 1999). Michele H. Bogart directed me to this text.

52. Schwab Albums. Less public rooms, according to the album catalogue lists, were given other subjects: for instance, a parlor received a pastoral love theme.

53. Ibid.; for additional images, see Hébert and *Architectural Record* 1902.

54. Hébert to Gérôme, Sept. 24, 1902; Gérôme to Hébert, Nov. 27, 1902; in Hébert, source of all correspondence between these two cited here. See also "Les Deux Statues de l'hôtel Schwab," *Gaulois* (December 31, 1902): 1, and C. L. B., "For Mr. Schwab's House," *New York Tribune*, 42, no. 20, 531 (Feb. 1, 1903), illustrated supplement: 5. My thanks to Pamela J. Warner for a photocopy of *Gaulois*. According to the above-cited C.L.B., Gérôme insisted instead that Schwab himself approached him for the commission while in Paris.

55. Gérôme to Hébert, Nov. 27, 1902.

56. *New York Tribune*.

57. A plaster in the Gérôme collection at the Musée Garret, Vesoul, may be the one surviving model for this project and is closer to the definitive version than the sketches in Fig. 5. See Gerald A. Ackerman, *Jean-Léon Gérôme: Monographie révisée. Catalogue raisonné mis à jour* (Courbevoie: ACR Éditions, 2000) 402, cat. S. 61, repro.; and *Jean-Léon Gérôme*...(Vesoul: Ville de Vesoul, 1981), 153, cat. no. 193. Peter Blume first linked this plaster, until then called a preliminary sketch for the *Grief* originally on the Gérôme family tomb (Montmartre Cemetery, Paris), with *Scientific Metallurgy*. See Ackerman's undated letter (received Nov. 16, 1981) to Blume; curatorial records, Allentown Art Museum.

58. Gérôme to Hébert, June 18, 1903.

59. Gérôme to Hébert, Dec. 30, 1903.

60. Creelman, 7.

61. Gérôme to Hébert, July 18, 1903. Prices for the models and the statuettes cited in various letters are taped over.

62. Gérôme to Hébert, June 18, 1903; Hébert to Gérôme, undated fragment (1903?). **Hébert claims the sculpture never arrived.**

63. Gérôme to Hébert, Dec. 30, 1903.

64. Possibly they went to "Immergrün," legally Rana's property, which her executors reportedly dismantled at her death.

65. Leaf 82, Hébert Folio (an oversize photographic album among the Schwab Albums that duplicates their content except for the photograph of *The Bather*). Though illustrated in the folio, it is unknown whether *The Bather* was actually placed at "Riverside." The figure had a pendant, a slightly smaller near-mirror-image. See Ackerman, 406–7, cat. nos. S. 73–73.2.

66. Note the relative scale of the figure to the room, tools, and what appears to be a boot resting on the workstand.

67. "Monsieur's Bathroom—East Elevation," leaf 32, Schwab Albums, 1983.148.460; see also Hébert, where the watercolor is entitled "Marble and mosaic bathroom."

68. The most extensive study of Gérôme's life and career remains Ackerman.

69. *New York Tribune*.

70. Charles Moreau-Vauthier, *Gérôme peintre et sculpteur: L'homme et l'artiste*... (Paris: Hachette, 1906), 263–66.

71. Paul Vitry, "Contemporary French Sculpture," *Architectural Record*, 12, no. 5 (Oct. 1902): 472–90.

72. In the fall of 1902, according to Gérôme, Hébert spent a month at Mathilde's country house; Gérôme to an unknown correspondent (Schwab?), Oct. 9, 1902. Schwab, in turn, claimed to have read every book in English about Napoleon (Hessen, 11).

73. Ackerman, 400–1, cat. no. S. 58.

74. Hébert to Gérôme, Sept. 24, 1902; Hébert, "Description of Architectural Styles of the Exterior and Decorative Styles of the Interior...," in Hébert. The *New York Tribune* erroneously reported that they were intended for the main staircase.

75. Creelman, 7.

76. Fig. 4, whose date is unknown, is evidence that the bronzes had not been transferred to the terrace by the time the photograph was taken. In the opinion of classic American car authority David Walters (personal communication), the motor vehicle in the photograph (the lowest of the three coaches at upper left) dates possibly around 1910–15, 1920 at the latest.

77. *Gaulois*; see the various press clippings in the Schwab United States Shipping Board albums, Bethlehem Steel Collection.

78. Hébert to Gérôme, Sept. 24, 1902; *New York Tribune*.

79. *New York Tribune*.

80. Creelman, 7.

81. Hébert to Gérôme, Sept. 24, 1902.

82. A good technical discussion of these processes is Henry Marion Howe, *Iron, Steel, and Other Alloys* (New York: McGraw-Hill, 1906), 343–430. See also John Howe Hall, "Manufacture of Bessemer Steel," and Kenneth Seaver, "Manufacture of Open-Hearth Steel," in A. O. Backert, ed., *ABC of Iron and Steel*, 5th ed. (Cleveland: Penton, 1925), 121–54.

83. Bulletin of the AISA; Grace, 8.

84. The press often criticized Schwab for disparaging higher education. See, for instance, E. S. Martin, "This Busy World," *Harper's Weekly*, 45, no. 2318 (May 25, 1901): 539.

85. *New York Tribune*.

86. Ibid.

87. *Gaulois*.

88. A classic discussion of the labor issues in the steel industry from the 1880s through the early twentieth century is David Brody, *Steelworkers in America: The Nonunion Era* (Urbana: University of Illinois, 1998). For a pro-management account of this period by a labor organizer, see James Davis, *Iron Puddler: My Life in the Rolling Mills and What Became of It* (New York: Grosset & Dunlap, 1922). For an economic history of the period, see Peter Temin, *Iron and Steel in Nineteenth-Century America: An Economic Inquiry* (Cambridge, MA: MIT, 1964). One of the most famous accounts of conditions specifically at the Pittsburgh plants during these years is a novel suggested to me by Lance Metz: Thomas Bell, *Out of This Furnace* (New York: Liberty Book Club, 1950).

89. Grace, 8–9, 10, 35.

90. *Gaulois*.

91. For a brief technical discussion of the process and its applications, see Hall, in Backert, 103–20.

92. I owe this idea to Lance Metz.

93. Davis, 92.

94. Ibid., 87.

95. Ibid., 90–113. For a more scientific account of the process, see James P. Roe, "Manufacture of Wrought Iron," in Backert, 93–102.

96. Howe, 346.

97. Roe, in Backert, 101–2.

98. Hessen, 257.

99. *Bethlehem Steel Company: Historical Sketch*, Bethlehem, PA Online! (www.bethlehempaonline.com/bethsteel.html), 2–3. The online article was extracted from the *Souvenir History Book of The Borough of South Bethlehem, Pennsylvania*, published in 1915.

100. Even artists exempt from the jury, like Gérôme, were limited to only two entries per exhibition by the governing body, the Société des Artistes Français. Gérôme's entries are cat. nos. 2922 (*Corinthe*) and 2923 (*Ouvrier Metallurgiste*). My thanks to Dominique Lobstein for a photocopy of the catalogue page.

101. There are several recorded versions of this subject, with slightly differing dimensions and surface treatments, in private collections where known. See Ackerman, 402–43, S. 63. For a discussion, see Ackerman, in Peter Fusco and H. W. Janson, eds., *Romantics to Rodin: French Nineteenth-Century Sculpture from North American Collections*, exh. cat., Los Angeles County Museum of Art, 1980, 291–92, cat. no. 157.

102. For a pictorial survey, see Maurice Rheims, *19th Century Sculpture*, trans. Robert E. Wolf (New York: Harry N. Abrams, 1977), 181–92. For discussion of French projects, see John M. Hunisak, "Images of Workers: From Genre Treatment and Heroic Nudity to the Monument to Labor, in Fusco and Janson, 52–60; and Philippe Durey, "Le réalisme," in *Sculpture française au XIXe siècle*, exh. cat., Galeries nationales du Grand Palais, Paris, 1986, esp. 362–71.

103. See Sura Levine and Françoise Urban, *Hommage à Constantin Meunier 1831–1905*, exh. cat. Galerie Maurice Tzwern, Antwerp, 1998; and Dabakis, 105–26.

104. Dabakis, 29–104. My thanks to Anne Pingeot for the photograph of the Meunier *Puddlers.*

105. Ibid.

106. Ibid., 83–104.

FREQUENTLY CITED SOURCES

AISA. American Iron and Steel Association, later known as the American Iron and Steel Institute (AISI.)

Schwab Albums. Seven of the eleven known albums, privately printed by architect Maurice Hébert in 1905, now in the Bethlehem Steel Collection, National Canal Museum, Easton, PA. They feature photographs of Hébert's watercolors of interior elevations, architectural and sculptural models, and construction views of the mansion he designed and built for Schwab in New York. Most volumes are identical in format and title: "On the Hudson: Residence of Chas. M. Schwab, Esq...." Despite some duplication, they complement, with more views and catalogue information, material published in Hébert (see below), a single-volume variant by the same title that instead contains photographs of montages made of press clippings, photographs, and Hébert's correspondence with artists hired for the project. Present in three known American libraries, including the Winterthur Library, examples of the latter edition were originally sent (according to internal evidence) to collaborating artists. Schwab's own deluxe version among the Schwab Albums is currently missing.

ISBN: 1-882011-52-X
Library of Congress: 2002107705

DESIGN
Natalie Salvucci and Phil Unetic

PRINTER
Printed in Iceland by Oddi Printing

PHOTOGRAPHY
Angelo Caggiano